Getting Started:

Info on Craft Foam, Adhesives, Skills, and More

Let's talk about what supplies and skills you'll need to be sure you achieve success the first time you start crafting with foam.

Foam comes in an endless array of color.

Craft Foam

Craft foam is usually sold in thicknesses like 2mm, 3mm, and 5mm. Thicker, sturdier sheets like the 5mm are great for the important structural parts of the bags and accessories, like the bag sides. The thinner, more flexible 2mm and 3mm sheets are what most people think of when they think of craft foam, and are great for flaps, decorations, straps, and virtually everything else. The projects in this book were created using all three thicknesses; the foam materials lists indicate 3mm foam, but you can interchange 3mm and 2mm as desired.

From left to right: sticky glue lines, sticky glue dots, double-sided adhesive tape, and double-sided adhesive sheets.

Adhesives

Adhesive choices are numerous and can be somewhat confusing. The reality is that there are certain stress points in these projects, and finding the correct adhesive ensures success. Stress points are the areas you want to pay close attention to and ensure good adhesion; stress points include the areas where flaps, straps, and sides are attached to the sides of the bag, as well as pocket corners. Some of the best adhesive options include double-sided adhesive tape (extra strong), double-sided adhesive sheets (extra strong), sticky glue dots, sticky glue lines, and a low temperature glue gun.

Using a low temperature glue gun makes assembly of the projects quite fast, and the glue is suitable for use in all steps of the projects, for stress points as well as embellishments. **However, even low temperature glue guns can cause burns, and young children should not be left alone with these tools.** While the glue coming out of the gun may not be considered hot, the nozzle on the glue gun can reach temperatures high enough to cause problems. The photographs in the book show the use of a low temperature glue gun for constructing all projects; I would suggest that if you choose to use a glue gun that you work together with the child, setting the glue gun out of the child's reach, applying the glue to the craft foam, and then letting the child attach the craft foam sections one at a time. **Warning: Read and follow all of the manufacturer's directions and guidelines before allowing children of any age access to this tool.**

There are many double-sided adhesive products on the market, but be sure to look for products that state they are suitable for adhesion to fabric, leather, and/or non-porous surfaces. Products used for paper crafting do not always have the amount of adhesion that is required for the projects in this book. This is especially important when dealing with high stress points such as flaps, sides, and pocket corners.

All the essential techniques for using double-sided tapes and sticky glue materials are shown on pages 4-5.

Double-Sided Adhesive Tape

1 Double-sided tape is great for attaching embellishments, as well as attaching pieces like pockets. Here, you can see using the tape for a pocket. The orange strip is the backing (usually paper or plastic). First, apply the tape to your pocket.

2 Next, peel off the backing.

3 Finally, stick the pocket onto your piece. You're done!

4 Double-sided tape is also good for decorative accents. Apply the tape to the accent piece, then peel of the backing.

5 Attach the accent piece to your project.

Double-Sided Adhesive Sheets

1 If you want to attach adhesive to larger pieces of foam or want to die-cut adhesive-backed pieces, use double-sided adhesive sheets.

2 Simply press the sheet onto your foam piece and peel off one side of the backing as you go. When you are ready to attach the foam somewhere, remove the remaining backing as you go.

Sticky Glue Lines

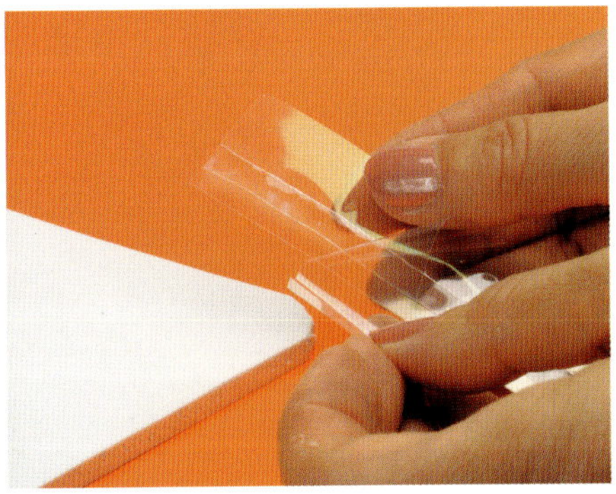

1 Sticky glue lines have a stronger hold than double-sided adhesive tape, making them perfect for stress points like the sides of the bags. Peel one backing layer off the glue line first.

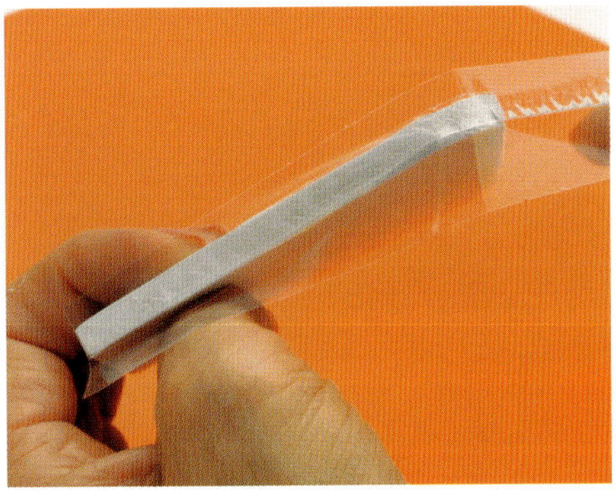

2 Stick the glue line onto the edges of your bag side.

3 Peel the other backing layer off the glue line.

4 Attach pieces to your bag side by pressing them onto the glue line.

5 This is a very simple, clean method.

Sticky Glue Dots

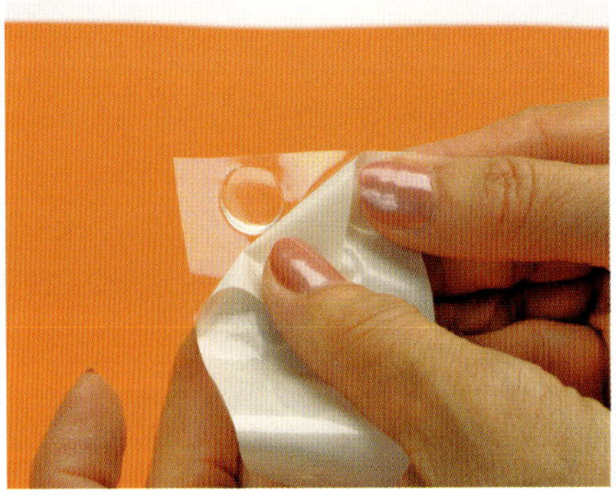

1 Glue dots are great for adding smaller embellishments and pom-poms. First, peel off one layer of backing.

2 Press the embellishment against the glue dot.

3 Pull the embellishment, along with the glue dot, off the remaining layer of backing, and attach it to your piece.

Decorative Accents & Embellishments

The options for embellishments are almost limitless! Check out the jewelry, paper crafting, sewing, and children's crafts sections in your local craft stores for ideas. Another option is to consider using repurposed, reclaimed, and recycled items. Here are just some of the items used for the projects in this book: permanent markers, plastic rings, beads, ribbons, plastic chain links, adhesive metallic dots, pre-cut craft foam stickers, die-cut craft foam accents (using a cutting machine and select dies), buckles, brads, and swivel clips.

There are several kinds of closures you can buy to complete your projects.

Use ribbons, brads, precut foam shapes, and more to embellish your projects.

Using a Die Cutter to Create Embellishments

You can add double-sided adhesive sheets to your foam pieces before cutting the shapes out for super easy stick-on embellishments. Below, learn how to use a simple crank handle die cutter to do so. *Note:* Follow the manufacturer's instructions for your specific machine. Some machines use different items for layering, or are computerized and run automatically; familiarize yourself with your machine before beginning.

1 First, set up the plate, your foam, and the die, layered as shown.

2 Add the second plate on top, creating a sort of sandwich.

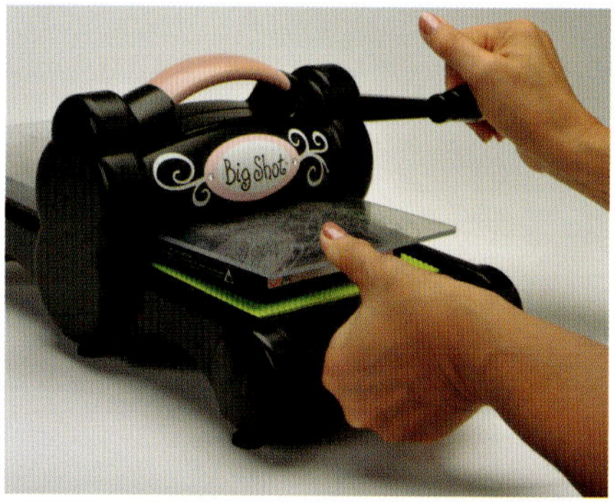

3 Feed the sandwich through the machine by turning the knob.

4 Here are your completed die cuts!

Patterns

The patterns provided in this book can be easily photocopied directly onto cardstock and cut out, or traced and then cut from cardboard to create a pattern tracer for children. Try gluing your pattern printouts onto cereal box cardboard and then cutting them out—that not only makes sturdy templates for children to trace around, but cereal boxes are fun and colorful, and you are reusing what would normally go in the trash!

Making Cereal Box Tracers

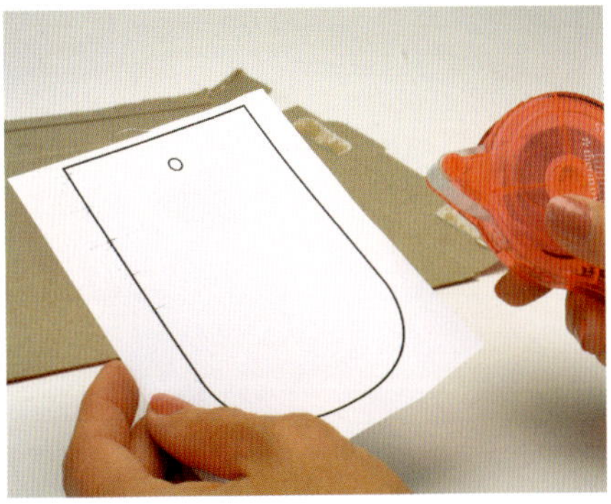

1 To make a sturdy pattern tracer from a cereal box, you need a photocopy of the pattern from the book, a tape roller, and your cereal box. When you cut out the paper pattern, you can leave extra space around it as shown, because you'll be cutting that off in step 3.

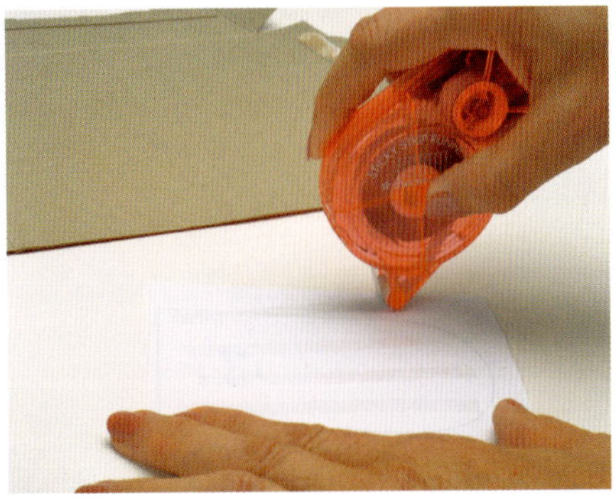

2 Roll tape along the back of the paper pattern in multiple strips. It's okay to go over the pattern edges.

3 Stick the pattern on the back of your cardboard and carefully cut the pattern out along the lines.

4 Use a hole punch to mark straps or similar spots.

Cutting

Cutting options are varied. A good pair of scissors is all that is absolutely necessary to cut out all of the pieces. When cutting the sturdy 5mm craft foam and/or cutting curves, you might find it simpler to use a template and a craft knife. If you are considering making these items with a large group and need to prepare numerous pieces beforehand, consider pulling out your rotary cutter, mat, and ruler. Craft foam cuts beautifully using these tools, and you will be done in no time at all. **Be careful not to allow children access to sharp craft knives.**

With just a few simple tools, many of which you already have at home, you can cut all the foam you need.

Especially for Little Ones!

Basic measuring, cutting, and gluing are all that is really needed to make every project in this book. Let the fun begin for the youngest children with embellishments. Older children can measure, cut, and glue their own purses and accessories from start to finish with a little supervision.

Small children will want to create their very own version of many of the projects in this book. Embellishments are the key to success. Purchase a variety of adhesive-backed pre-cut craft foam shapes, plastic jewels, ribbons, and beads. Set the stage by displaying the embellishments in dishes on the work surface alongside a finished sample of the project. Take a moment to explain how the children might use the embellishments, and then let them have at it! Depending on the project children are working on, give them one piece at a time to embellish. Try starting with a pocket to give them a chance to experiment. Depending on the type of adhesive you are using, most items are removable if taken off within minutes of their placement. Pass out additional sections of the projects to be embellished as the children work. Once all the pieces are decorated, assemble their projects quickly and easily using a glue gun while they continue with another game, activity, or snack. The projects in this book are perfect for parties for so many reasons: the projects, as well as the embellishments, can be themed, they are a fun crafting activity for groups, and the finished projects can be used as goodie bags by slipping some inexpensive toys or candies into the completed bags.

Useful Tips & Techniques

Here are four additional techniques that will make it so much easier to make lovely, well-proportioned, and creative bags. Don't skip this section!

Lining Up Hook-and-Loop Adhesive Dots

No guessing necessary when you follow these instructions every time!

1 After attaching one half of the hook-and-loop adhesive dot to your bag flap, take the other half and place it on top of its partner.

2 Fold the flap down to close the bag and press the hook-and-loop closure into the bag body.

3 The other half of the dot will stick to the bag body, aligned exactly where you want it.

Using a Roller

This tip will ensure your bags look squeaky clean.

You can use a roller to press two glued pieces together to achieve nice, smooth adhesion.

Interchangeable Embellishments

One of the most fun things about making your own foam accessory is that you can customize its design any way you want. But what's even better than that? Being able to customize your item over and over again! With just a few simple steps, you'll be able to make embellishments that are removable, so you can change decorations in and out depending on your mood or style. That way, you won't ever have to worry about making the perfect piece, because you'll be able to change your piece any way you want. Try it!

1 Apply one half of a hook-and-loop adhesive dot to the place you want to create an interchangeable embellishment. Here, it is on the front of the bag flap.

2 Apply the other half of the hook-and-loop adhesive dot to the back side of your embellishment. You won't see the embellishment's hook-and-loop dot while it's attached, so it can be any color; it doesn't have to match the bag or even the other half of the dot!

3 Stick the embellishment onto your bag, and you're done.

4 Create as many removable embellishments as you like by layering pre-cut foam shapes, cutting your own new shapes, and selecting pom-poms and jewelry. You can change your bag's look whenever you want.

Using Brads

Here is the basic technique for applying brads to a project. When using brads, be careful of the sharp ends.

1 Poke the sharp end of the brad through your foam pieces.

2 If you are attaching one piece to another, as in the strap shown, insert the end of the brad through a pre-made hole.

3 Flatten the ends of the brad using an awl or similar tool.

Purse/Tool Bag

The first bag you can make uses a lot of the techniques that you'll use for the other bags. You can make a lovely purse to carry around makeup, accessories, or other goodies. Or make a handy tool bag decorated with hammers and screws. You'll want to take your new bag everywhere!

Supplies

- Decorative brads
- Adhesive-backed foam shapes
- Hook-and-loop adhesive dots
- Large needle or awl

Foam Cutting Chart

3mm Pieces	Color	Number to Cut	Size
Bag Body	Lime	1	6" x 12" (15 x 30cm)
Flap	Pink	1	5" x 6" (12.5 x 15cm)
Strap	Lime	1	1" x 12" (2.5 x 30cm)
Strap Accent Strip	Pink	1	½" x 12" (1.5 x 30cm)
Inside Reinforcement Strip	Pink	1	1" x 6" (2.5 x 15cm)
Flap Accent Strip	White	1	1" x 6" (2.5 x 15cm)
5mm Pieces	**Color**	**Number to Cut**	**Size**
Bag Side	White	2	Pattern provided

1 Collect all materials and cut out all foam pieces according to the cutting chart.

2 Apply adhesive to the side and bottom edges of one bag side piece at a time.

3 Working in 2" (5cm) sections, press the long edge of the bag body into the adhesive, wrapping it around three edges of one bag side piece at a time. Hold until the adhesive sets. The bag body will extend about 1½" (4cm) past the bag sides at the back (see the photo for step 5).

4 Apply adhesive to the flap accent strip, and attach it to the bottom front edge of the flap.

5 Apply adhesive to the underside of the flap on the opposite side from the accent strip, and attach it to the bag body, overlapping the two edges 1½" (4cm) for support. You can also use brads to attach the flap if desired.

6 Apply adhesive to the inside reinforcement strip, and attach it to the inside of the bag body, even with the top edge of the back of the bag as shown. (If you choose to use brads to attach the flap, this will cover up their ends. Read more about using brads on page 9)

7 Apply adhesive to the back of the strap accent strip, and attach it to the center of the strap.

8 Using an awl or large needle, poke a hole through each bag side at the position marked on the bag side pattern. You can mark it with pen first, because the hole and the pen mark will be covered up. Children should get help from an adult for this step.

9 Insert brads through the ends of completed strap and then through the holes in the bag sides, and open the ends of the brads. The strap foam should be thin enough that you can push the brad through without punching a hole.

10 Cover the brad ends on both sides with a 1" (2.5cm)-diameter adhesive-backed circle.

11 Apply one half of a hook-and-loop adhesive dot to the underside of the flap and the other half to the front of the bag body for the closure.

12 Embellish the flap as desired.

Clutch/Pencil Case

Whether you make a vibrantly patterned clutch that is perfect for dress-up, or a cool geometric personalized pencil case, everyone will be amazed at your unique style.

Supplies

- Swivel clip
- D-ring
- Hook-and-loop adhesive dots

Foam Cutting Chart

3mm Pieces	Color	Number to Cut	Size
Clutch Body	Print	1	12" x 7" (30 x 18cm)
Inside Pocket	Solid	1	3" x 4" (7.5 x 10cm)
Side Accent Strip	Print	2	3/4" x 12" (2 x 30cm)
Inside Pocket Accent Strip	Print	1	1/4" x 4" (0.5 x 10cm)
Strap	Print	1	3/4" x 12" (2 x 30cm)
Strap Accent Strip	Solid	1	1/4" x 11" (0.5 x 28cm)
5mm Pieces	**Color**	**Number to Cut**	**Size**
Clutch Side	Black	2	1" x 3 1/2" (2.5 x 9cm)

1 Collect all materials and cut out all foam pieces according to the cutting chart.

2 Apply adhesive to the side and bottom edges of one clutch side piece at a time. Working in 2" (5cm) sections, press the long edges of the clutch body into the adhesive, wrapping it around the three edges of one clutch side piece at a time. The bag body will extend past the bag sides at the back to form the flap (you can see this in the left side of this photo). Hold until the adhesive sets.

3 Apply adhesive to the inside pocket accent strip, and attach it to the front top edge of the inside pocket.

4 Apply adhesive to the back of the inside pocket along the sides and bottom.

5 Attach the pocket to the inside of the clutch as shown, centering it ½" (1.5cm) down from the top of the side pieces of the clutch.

6 Slide the D-ring onto one side accent strip, fold the strip over 1½" (4cm), and glue the folded piece down.

7 Apply adhesive to one clutch side and press the side accent strip into the adhesive, letting the end with the D-ring extend 1" (2.5cm) above the top edge of the clutch. Continue adding adhesive and pressing the side accent strip into the clutch side, working in 2" (5cm) sections at a time.

8 Continue attaching the side accent strip along the bottom of the clutch, adding more adhesive as you go until you reach the end of the strip.

9 Butt the end of the second side accent strip up to the end of the first strip and continue adding adhesive and pressing it in place around the bottom of the clutch and up the opposite side.

10 Trim away any excess from the second side accent strip even with the top edge of the clutch side.

11 Slide the swivel clip onto the strap, fold the strap over 1" (2.5cm), and glue the folded piece down.

12 Apply adhesive directly below the folded and glued end of the strap, bring the opposite end up into a loop, and press it down onto the adhesive, aligning it with the edge of the folded portion.

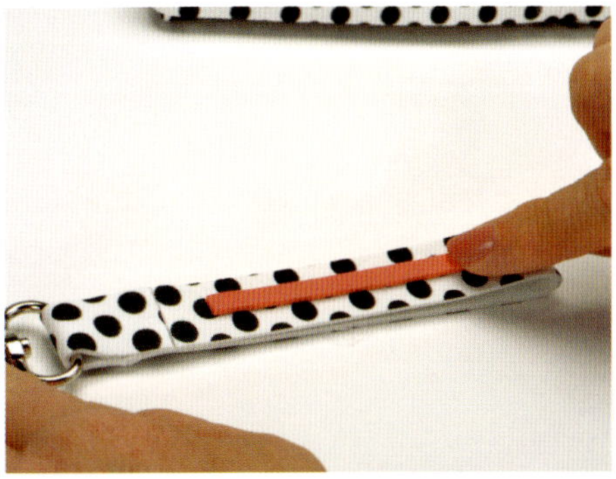

13 Apply adhesive to the strap accent strip and attach it to the center of the strap, wrapping it around the entire strap until the ends meet.

14 Clip the strap onto the body of the clutch.

15 Apply hook-and-loop adhesive dots for the closure, and embellish with foam shapes and adhesive accents as desired.

Foam Charm

Decorate a backpack, lunch bag, or use it as a zipper pull. All you need is some foam and some ribbon.

Supplies

- Adhesive-backed foam shapes
- 10" (25.5cm) of ⅛" (0.25cm)-wide satin ribbon
- One pom-pom

1 Remove the paper backing from one foam shape and, with the ribbon folded in half, press the ribbon into the adhesive so that the ribbon loop extends beyond the top of the shape. You can leave a decorative tail by letting the two loose ribbon ends stick out, or hide them inside the shape.

2 Remove the paper backing from a second adhesive foam shape and press it onto the back of the first shape, sandwiching the ribbon.

3 Apply adhesive to the back of a pom-pom and press it in the center of the front of the zipper pull. You can add other embellishments, too (like the extra white flower in the photo). To attach your zipper pull, simply loop the ribbon through and tie it to the zipper.

Messenger Bag: Style 1

The ultimate handy accessory for any student or adventurer is a large messenger bag with a long strap that can carry whatever you need. Mix prints for funky combinations, decorate with patriotic stars–the possibilities are endless.

Supplies

- Adhesive-backed metallic accents
- Plastic chain link
- Fine tip permanent marker

Foam cutting chart

3mm Pieces	Color	Number to Cut	Size
Front Flap	Print	1	Pattern provided
Bag Body	Yellow	1	7½" x 11½" (19 x 29cm)
Front Accent Strip	Print	1	1" x 7½" (2.5 x 19cm)
Strap Loops	Yellow	2	½" x 8" (1.5 x 20.5cm)
Side Pocket	Print	1	3" x 2½" (7.5 x 6.5cm)
Front Fastener Tab	Yellow	1	1" x 8" (2.5 x 20.5cm)
Inside Pocket	Print	1	3" x 4½" (7.5 x 11.5cm)
Front Fastener Loop	Yellow	1	1" x 4" (2.5 x 10cm)
Inside Pocket Accent Strip	Yellow	1	¼" x 4½" (0.5 x 11.5cm)
5mm Pieces	**Color**	**Number to Cut**	**Size**
Bag Side	White	2	Pattern provided

Make the gingham print variation on page 21!

1 Collect all materials and cut out all foam pieces according to the cutting chart.

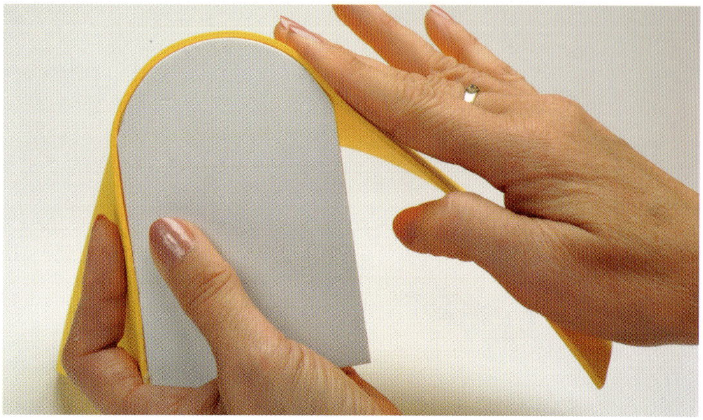

2 Apply adhesive to the side and bottom edges of one side piece. Working in 2" (5cm) sections, press one long edge of the bag body into the adhesive, wrapping it around the three edges of the side piece. Hold until the adhesive sets.

3 Repeat for the opposite side of the bag. When finished, the bag body will extend past the top of the bag sides at the back (you can see this in the background of this photo).

4 Apply adhesive to the inside pocket accent strip. Attach the strip to the front of the inside pocket ¼" (0.5cm) down from the top edge.

5 Apply adhesive to the back of the inside pocket along the sides and bottom. Attach the pocket to the inside of the bag body, centering it 1" (2.5cm) down from the top edge.

6 Apply adhesive to the underside of the front flap along the straight edge, and attach the flap to the top edge of the bag body that extends past the sides of the bag, overlapping by 1" (2.5cm) for support.

7 Apply adhesive to the front accent strip. Attach the strip to the front of the bag body, even with the top edge.

8 Apply adhesive to the back of the front fastener tab and fold the ends in toward the center so the edges meet. Press it flat. This makes the tab double-layered for extra strength.

9 Add decorative stitching lines to the tab with a fine tip permanent marker.

10 Attach adhesive-backed metallic accents to the tab.

11 Apply adhesive to about half of the back of the tab and attach it to the center front of the front flap, 2" (5cm) up from the bottom edge.

12 Apply adhesive to each end of the front fastener loop and attach it to the front of the bag, centering it $1\frac{1}{2}$" (4cm) below the front accent strip. It is helpful to hold the flap with the tab down so that you can glue the fastener loop over the tab and align it correctly.

13 Check to make sure you can slide the tab in and out of the loop without too much trouble.

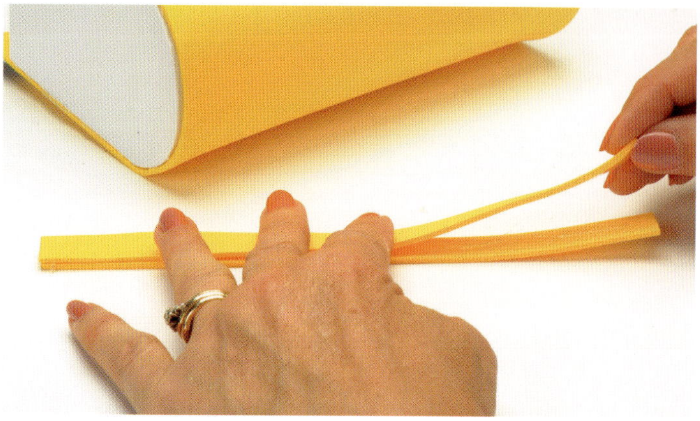

14 Apply adhesive to one of the strap loops and attach it the other to make a double-layered single piece. Then cut the piece in half so that you have two 4" (10cm)-long pieces.

15 Fold one 4" (10cm) piece into a loop as shown. If you are using a chain or other material that can't be opened and closed, slide the strap onto the loop prior to gluing it to the bag. Apply adhesive to the bottom of each side of the loop and attach it to one side of the bag, about 2" (5cm) down from the top edge.

16 Apply adhesive to the back of the side pocket along the sides and bottom and glue it to the bag side, over the strap loops and ½" (1.5cm) down from the top edge.

17 Attach the chain to the strap loop.

18 Use the other 4" (10cm) piece from step 14 to make the loop on the other side of the bag. Here you can see the other side of the bag, without a pocket.

Messenger Bag: Style 2

This version of the messenger bag is put together in much the same way as the first, but has some important decorative and functional differences. Make both, or mix and match features from each bag to create a whole new style!

Supplies

- Adhesive-backed metallic accents
- Adhesive-backed foam shapes
- Hook-and-loop adhesive dots

Foam cutting chart

3mm Pieces	Color	Number to Cut	Size
Bag Body	Black	1	7½" x 11½" (19 x 29cm)
Front Flap	White	1	7½" x 5" (19 x 12.5cm)
Straps	Yellow	2	1½" x 18" (4 x 45.5cm)
Strap Accent Strips	Yellow	4	¼" x 18" (0.5 x 45.5cm)
Strap Reinforcement	Yellow	1	1½" x 2" (4 x 5cm)
Bag Accent Strips	Yellow	2	¼" x 11½" (0.5 x 29cm)
Inside Bag Reinforce-ment Strip	White	1	1½" x 16" (4 x 40.5cm)
Inside Pocket	Yellow	1	2½" x 3" (6.5 x 7.5cm)
5mm Pieces	**Color**	**Number to Cut**	**Size**
Bag Side	White	2	Pattern provided

Get creative to make this adventure bag. The general bag shape is the same, but the embellishments are different.

1 Collect all materials and cut out all foam pieces according to the cutting chart.

2 Apply adhesive to the bag accent strips and attach them to the long edges of the bag body, ⅛" (0.25cm) in along each side.

3 Apply adhesive to the back of the inside pocket along the sides and bottom, and attach it to one of the bag sides, centering the pocket ½" (1.5cm) down from the top edge.

4 Apply adhesive to the side and bottom edges of one bag side piece. Working in 2" (5cm) sections, press one long edge of the bag body into the adhesive, wrapping it around the three edges of the side piece. Hold until the adhesive sets.

5 Repeat for the opposite side of the bag. When finished, the bag body will extend ½" (1.5cm) past the top of the bag sides at the back. Be sure to attach the bag side that has the pocket so that the pocket is inside of the bag.

6 Apply adhesive to the underside of the front flap along one of the long edges and attach it to the top edge of the bag body that extends past the sides of the bag, overlapping by ½" (1.5cm) for support.

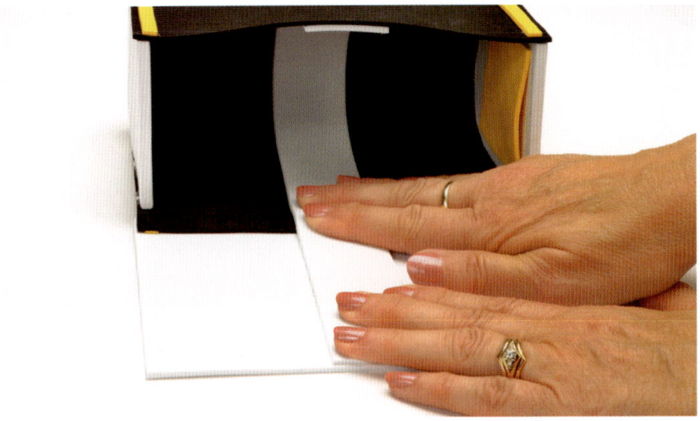

7 Apply adhesive to the bag reinforcement strip and attach it to the inside of the bag, centering the strip as shown. If gluing, do only half of the strip at a time, starting by applying glue to the part of the strip on the inside of the bag, pressing it down, and then applying glue on the part of the strip that attaches to the flap.

8 Butt the two straps together end to end. Then, apply adhesive to the strap reinforcement and attach it over the place where the two strap pieces touch. This connects them to create one long strap.

9 Cut two of the strap accent strips in half, so that you end up with four short strips and two long strips (uncut). Apply adhesive to one short strip, align one end with one end of the strap, and attach it along the long edge. Apply adhesive to a long strip, butt it against the short strip you just attached, and attach it. Then, attach another short strip butted against the other end of the long strip. This will complete an accent strip that runs the length of the entire strap. Repeat with the remaining strips for the other side of the strap. This cutting method is used to avoid a seam on top of a seam at the center of the strap.

10 Apply adhesive to the underside of the ends of the strap and attach them to the bag sides.

11 Apply three hook-and-loop adhesive dots to the bag flap and the body of the bag for the closure.

12 Embellish with foam shapes and metallic accents.

Tassel

A colorful tassel will jazz up anything that needs a little pizzazz. It may look tricky to make, but it's really quite simple.

Supplies

- 6" (15cm) of 1/4" (0.5cm)-wide satin ribbon
- Plastic beads

Foam Cutting Chart

3mm Pieces	Color	Number to Cut	Size
Tassel	Yellow	1	1½" x 5" (4 x 12.5cm)
Accent Strip	Print	1	¼" x 2½" (0.5 x 6.5cm)

1 Collect all materials and cut out all foam pieces according to the cutting chart.

2 Cut into the long edge of tassel piece with repeated cuts ¼" (0.5cm) apart, cutting along the entire 5" (12.5cm) length. Leave a ¼" (0.5cm) strip intact along the top edge.

3 Apply adhesive all along the uncut top edge of the tassel, fold the ribbon in half, and press it into the adhesive so the loop of the ribbon extends 1" (2.5cm) down from the glue.

4 Start rolling the tassel, beginning at the end with the ribbon. The two loose ribbon ends stick out of the top of the tassel so that it can be tied on to anything.

5 Apply adhesive to the accent strip and wrap it around the top of the tassel, trimming away any excess.

6 Slide several beads onto the two ribbon ends, and add some dots of glue to the ribbon at the base of the tassel. Slide the beads down over top of the glue. This will keep the beads together and keep them down against the tassel.

Key Fob

This key fob can be used for keys or just attached to a backpack or other accessory as decoration. Use your initial to show the world it is your creation.

Supplies

- Adhesive-backed foam letter
- 4" (10cm) of 3/8" (1cm)-wide grosgrain ribbon
- Plastic swivel clip
- Key ring
- Fine tip permanent marker

Foam Cutting Chart

3mm Pieces	Color	Number to Cut	Size
Circle	Blue	1	Pattern provided
Circle	Yellow (adhesive-backed*)	1	Pattern provided

*Apply double-sided adhesive to the circle before cutting it out, if necessary.

1 Collect all materials and cut out all foam pieces according to the cutting chart.

2 Remove the paper backing from the foam letter and press it in place on top of the yellow adhesive-backed circle.

3 Attach the swivel clip to the key ring. Slide the key ring onto the ribbon and fold the ribbon in half to form a loop. Remove the paper backing from the yellow circle and press the ribbon ends into the adhesive, allowing the top loop to extend a bit beyond the edge of the circle.

4 Cover the back of the yellow circle with the blue circle, pressing them together to sandwich the ribbon.

5 Use a fine tip permanent marker to add stitching lines.

Wallet

Just three pieces of foam and some decorations come together to make a perfectly functional wallet. Don't spend your money all in one place!

Supplies

- Adhesive-backed foam letter
- Hook-and-loop adhesive dots
- Fine tip permanent marker

Foam Cutting Chart

3mm Pieces	Color	Number to Cut	Size
Wallet Body	Blue	1	4" x 10" (10 x 25.5cm)
Large Inside Pocket	Blue	1	2½" x 8" (6.5 x 20.5cm)
Small Inside Pocket	Blue	1	2" x 3" (5 x 7.5cm)

1 Collect all materials and cut out all foam pieces according to the cutting chart.

2 Apply adhesive to the back of the large inside pocket along the sides and bottom, as well as in a line in the center. Attach the pocket to the wallet body so it's even with the bottom left corner. In this photo you can see some double-sided adhesive tape on the foam piece, which is another option for adhesion.

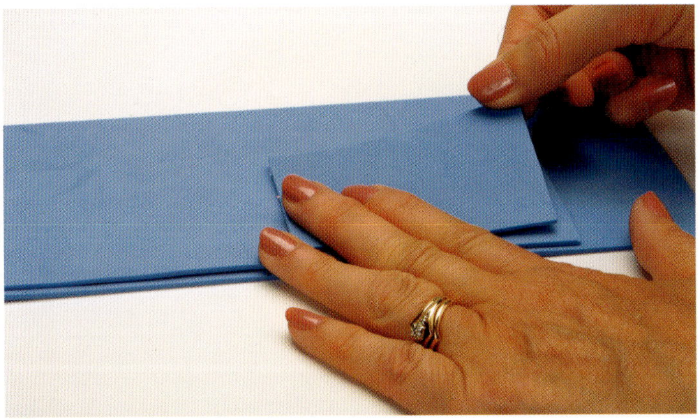

3 Apply adhesive to the back of the small inside pocket along the sides and bottom. Attach it on top of the large pocket as shown.

4 Add hook-and-loop adhesive dots to the wallet for the closure.

5 You can decide how you want the wallet to fold depending on the placement of the dots. Here, the wallet folds over onto the outside front. In the photo on page 26, the wallet folds on the inside so that you don't see a fold on the front of the wallet.

6 Embellish the outside of the wallet with an adhesive-backed foam letter, and add stitching lines to the letter and wallet using a fine tip permanent marker.

Craft Carryall

This funky modern design is perfect for carrying all the craft items you need to get creative, from small drawings and pens and pencils to scissors and other helpful tools. With a handy inner pocket and a secure strap, tote your supplies in style.

Supplies

- Plastic buckle closure
- Eight 1/2" (1.5cm)-diameter plastic rings
- Hook-and-loop adhesive dots
- 1/4" (0.5cm)-diameter hole punch

Foam Cutting Chart

3mm Pieces	Color	Number to Cut	Size
Bag Body	Purple	1	8" x 15" (20.5 x 38cm)
Side Pockets	Assorted textured	2	6" x 5" (15 x 12.5cm)
Handles	Purple	2	1/2" x 15" (1.5 x 38cm)
Center Pocket	Assorted textured	2	6" x 8" (15 x 20.5cm)
Buckle Strap	Purple	1	1/2" x 18" (1.5 x 45.5cm)
Center Spine	Assorted textured	1	1" x 8" (2.5 x 20.5cm)
Center Pocket Tab	Purple	1	1 1/2" x 3 1/2" (4 x 9cm)
Center Spine Reinforcement Strip	Assorted textured	1	3/4" x 8" (2 x 20.5cm)

1 Collect all materials and cut out all foam pieces according to the cutting chart.

2 Mark the center line of the bag body (at $7\frac{1}{2}$" [19cm]).

3 Apply adhesive to the center spine reinforcement strip and attach it in the center of the bag body, over the line you drew in step 1.

4 Apply adhesive to the two long sides of one of the center pocket pieces and attach the second pocket piece on top.

5 There will be an opening through the middle.

6 Center the pocket over the center spine. Apply adhesive to the center spine and press it over the pocket and center spine reinforcement strip.

7 Apply adhesive to one end of the center pocket tab and attach it to the center back of the left side of the pocket.

8 Apply hook-and-loop adhesive dots to the end of the tab and the back of the right side of the pocket.

9 The pocket will close as shown.

10 Apply adhesive to the back of the two side pockets along the sides and bottom, and glue them to the inside of the bag body, one on each side of the center pocket. See the photos for placement.

11 Mark the placement for the handles at four spots by measuring 1" (2.5cm) in from the bag's short sides and 1½" (4cm) in from each of the bag's long sides. Use a hole punch to create four holes at the marked spots, two on each side of the bag.

12 Slide a plastic ring onto a handle piece, insert the handle into one of the holes made in step 10, slide a second ring onto the other side, and push the rings down next to the hole.

13 Then, fold the end of the handle over 1½" (4cm) and glue it onto itself. Do this on the other end of the handle with the hole on the same side of the bag. Repeat with the remaining handle and two holes on the other side of the bag.

14 The rings reinforce the handle holes.

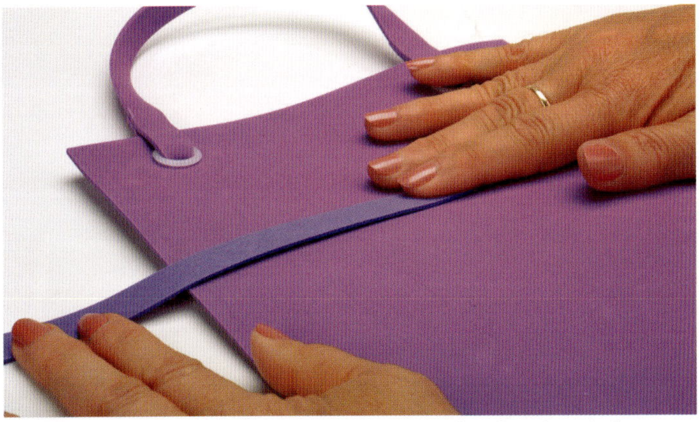

15 Center the buckle strap on the back of the bag so that equal lengths extend past the sides of the bag, and place it about 3" (7.5cm) down from the top edge of the bag. Apply adhesive to attach it.

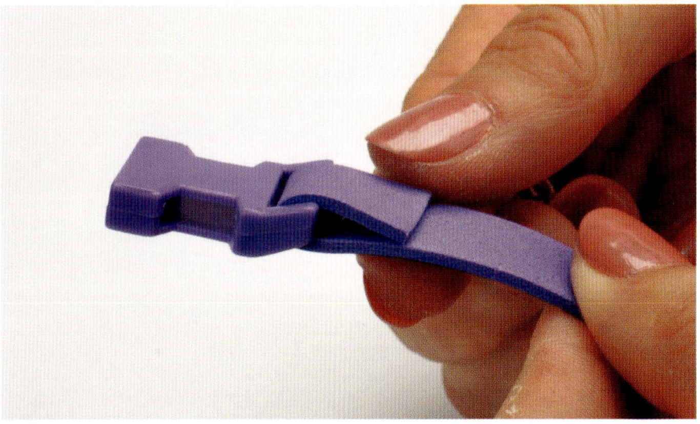

16 Slide one half of the buckle closure on one end of the buckle strap, fold the strap over about 1" (2.5cm), and glue it down.

17 Take the other half of the buckle and clip the two buckle pieces together, then loop the other end of the strap through the other half of the buckle. Adjust the length of the strap to fit comfortably around the bag while the buckle is closed, fold over the end of the strap, and glue it down.

18 Embellish with random cuts from textured craft foam, foam shapes, or other accents as desired.

Belt Pocket

Sometimes you need easy access to those important things and just don't have pockets in your jeans! So here is a sturdy and stylish contraption that you slide onto your belt. Nifty!

Supplies

- Decorative brads
- Hook-and-loop adhesive dots

Foam Cutting Chart

3mm Pieces	Color	Number to Cut	Size
Pocket Front	Blue textured	1	3½" x 6" (9 x 15cm)
Pocket Body	Yellow	1	6" x 10" (15 x 25.5cm)
Pocket Straps	Blue textured	2	1" x 8½" (2.5 x 21.5cm)
Belt Loop Reinforcement Strip	Yellow	2	1" x 2" (2.5 x 5cm)
Inside Reinforcement Strip	Blue textured	1	½" x 5½" (1.5 x 14cm)
5mm Pieces	**Color**	**Number to Cut**	**Size**
Pocket Side	White	2	½" x 3½" (1.5 x 9cm)

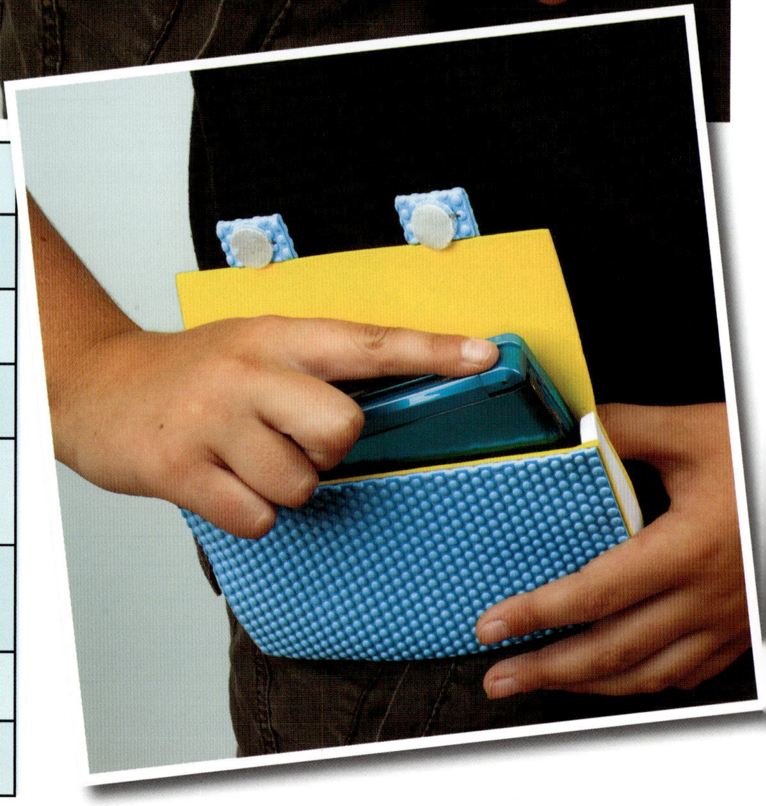

1 Collect all materials and cut out all foam pieces according to the cutting chart.

2 Insert a brad into a pocket strap, 1" (2.5cm) down from one end. Open the ends of the brad.

3 Apply adhesive to the end of the strap and fold it over, covering the end of brad. Repeat steps 1 and 2 for the second pocket strap.

4 Apply adhesive to the belt loop reinforcement strips and attach them to the back of the pocket straps, placing them 4" (10cm) from the folded ends.

5 Apply adhesive to the back of the pocket straps, but not to the belt loop reinforcements. These are not glued down because they are where the belt will slide through.

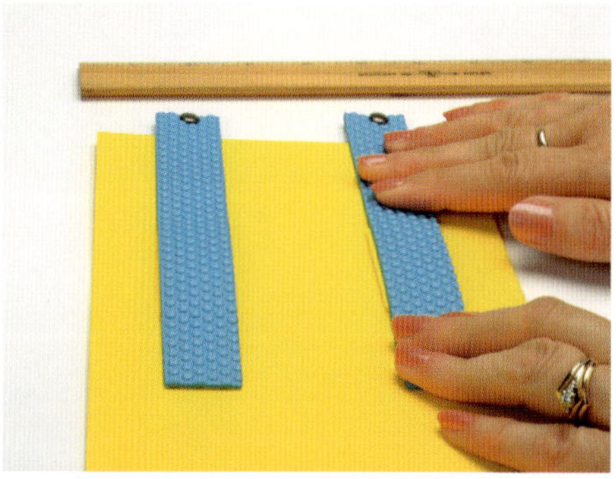

6 Press the straps in place on top of the pocket body, letting the top fold of each strap extend past the top edge of the pocket body. Place the straps 1" (2.5cm) in from each long side of the pocket body.

7 Insert brads into the bottom ends of the pocket straps and through the pocket body.

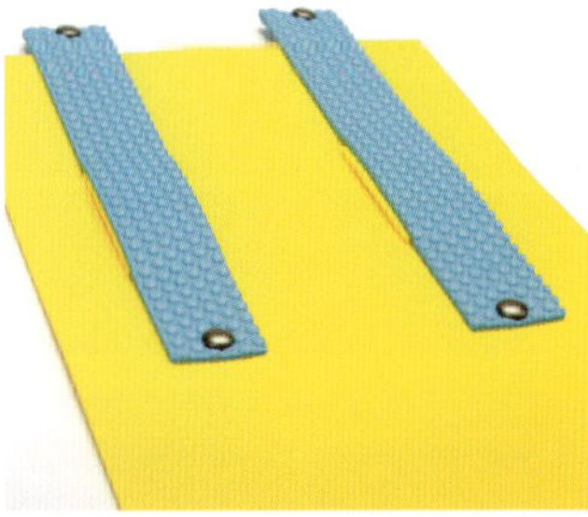

8 Here you can see all four brads.

9 Apply adhesive to the side and bottom edges of one pocket side piece at a time and, working in 2" (5cm) sections, press the long edges of the pocket body into the adhesive, wrapping it around the three sides of one side piece at a time. The pocket body will extend past the sides at the back to form the flap (you can see this in this photo). Hold until the adhesive sets.

10 Apply adhesive to the inside reinforcement strip, and insert it inside the pocket to cover the brad ends along the bottom of the pocket.

11 Apply adhesive to the pocket front and press it to the front of the pocket body.

12 Apply hook-and-loop adhesive dots to the straps for the closure.

Patterns

All patterns shown Full Size

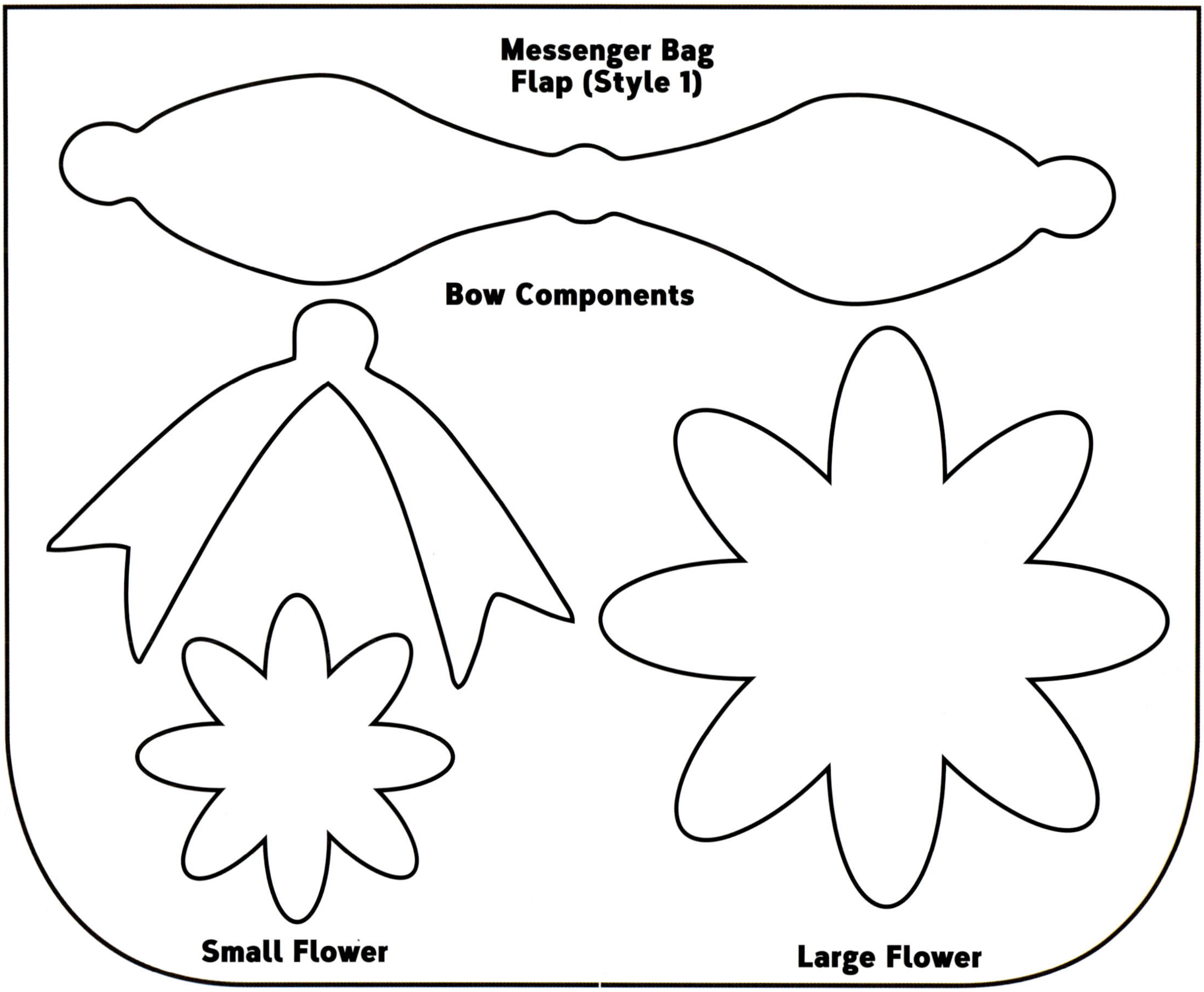